This book
belongs to

Sophie

CD-ROM
FACT*finders*
INTERACTIVE MULTIMEDIA
THE EARTH

Written by
Roger Coote

Consultant
Keith Lye

Edited by
Paul Harrison

zigzag
Multimedia

The author, Roger Coote, studied geography at Reading University and has written over 50 books for children.

ZIGZAG PUBLISHING

Published by Zigzag Publishing,
a division of Quadrillion Publishing Ltd.,
Godalming Business Centre, Woolsack Way, Godalming,
Surrey GU7 1XW, England

Series concept: Tony Potter
Senior Editor: Nicola Wright
Design Manager: Kate Buxton
Production: Zoë Fawcett
Designed by: David Anstey
Illustrated by: Peter Bull, Mainline Design, Kridon Panteli,
Steven Young, Michael Saunders and
Jeremy Gower
Cover design: Clare Harris

Colour separations: Scan Trans, Singapore
Printed in Singapore

Distributed in the U.S. by SMITHMARK PUBLISHERS
a division of U.S. Media Holdings, Inc.,
16 East 32nd Street, New York, NY 10016

ISBN 0-7651-9348-5
8409

Contents

Have you ever wondered why volcanoes erupt, or where the coldest place on Earth is? This book will answer these and other fascinating questions about our planet, Earth.

Learn about oceans, rivers, and climate and see how the Earth has changed through the ages. You will find out what is under the ground and what is on the ocean floor. Discover why cities are warmer than the countryside and what causes acid rain.

Packed with facts and full of colorful illustrations, this book clearly answers all of your questions about Earth.

What is a planet?

The Earth weighs about 6,700,000,000,000, 000,000,000 tons.

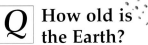

A planet is a world that travels around, or orbits, a star. Our planet - the Earth - orbits a star which we call the Sun. Both the Sun and the Earth are part of the Solar System.

Q What is the Solar System?

A It is a group of planets and about sixty moons in orbit around the Sun. The planets are Mercury, Venus, Earth, Mars, Jupiter, Saturn, Uranus, Neptune, and Pluto. Two newly discovered planets are called Smiley and Karla. There are also thousands of lumps of rock, called asteroids, which also orbit the Sun.

Q How was the Solar System formed?

A The Sun and its planets formed from a cloud of gas and dust whirling in space. The cloud was pulled together by gravity and it became very thick. Most of the cloud turned into what is now the Sun, and the rest made the planets, moons and asteroids.

Q Are there any other planets in space?

A The Solar System is a tiny part of a galaxy - the Milky Way - which contains 100 billion stars. The Milky Way is just one of more than ten billion galaxies in the Universe. It is almost certain that some other stars have planets, but they are hard to find because they are so far away. Scientists think they have found one about the same size as the Earth- it is about 186 million billion miles away.

Callisto

Europa

Jupiter

Asteroids

Moon   Earth

Mars

Sun

Venus

lo

Ganymede

Mercury

Q How old is the Earth?

A Scientists think that the Universe began than fourteen billion years ago. The Earth and the rest of the Solar System are much younger - more than four billion years old.

The Milky Way is a good example of a spiral galaxy. It measures almost 70,000 light years across.

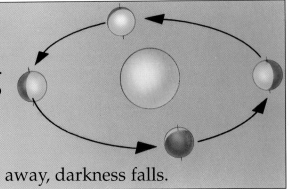

Q Why do we have day and night?

A The Earth gets light and heat from the Sun. As the Earth travels around the Sun, it also spins on its own axis (an imaginary line joining the North Pole, the center of the Earth and the South Pole). When your part of the Earth is facing the Sun you have daylight. When it turns away, darkness falls.

Saturn

Titan

Uranus

Neptune

Smiley

Karla

Pluto

Charon

Q What is the right time?

A It depends where you are, because when it's day in one place it's night somewhere else. There are 24 time zones around the world. The time is different in each zone so you must reset your watch when you travel from one zone to another.

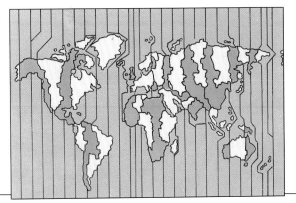

Q What shape is the Earth?

A Our planet is almost a round sphere - like a ball - but not quite. The speed at which the Earth spins causes the planet to bulge at the Equator and flatten at the Poles. This shape is called a spheroid.

Q If the Earth is spinning, why don't we fly off into space?

A We are held down by gravity. Everything "pulls" on everything else with a force - gravity. For most things gravity is too weak to notice, but the Earth is so large that it has a strong pull of gravity toward its center. The Sun's gravity is even stronger and it holds all the planets in orbit.

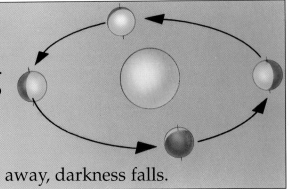

The temperature of the Earth's inner core is about 15,574°F.

What is the Earth made of?

The plates that make up the outer layers of the Earth move between 0.4 and 4 in. each year.

No one has dug down deep enough to be certain, but we think that it is made of layers of rock and metal. Scientists have worked this out by studying earthquakes.

Q What is inside the Earth?

A There are four main layers. The outer one is a layer of solid rock called the crust. Below this is the mantle. This is solid at the top, but deeper down it is so hot that the rocks have melted. Next is the outer core, made of hot liquid metal. At the center is the inner core, which is solid metal.

Q Could we dig a tunnel to the other side of the Earth?

A No. The inner core of the Earth is far too hot. The deepest hole dug into the Earth is about 7.4 mi. deep.

Q Is the Earth magnetic?

A Yes, the Earth acts as though it has a huge bar magnet inside it, with a magnetic field and north and south poles. A compass needle points to the north magnetic pole. The Earth's magnetic field stretches thousands of miles into space.

Q What is soil made of?

A There are many types of soil. They are all made of a mixture of pieces of rock, living particles, dead plant and animal material, air and water.

Crust

Mantle

Outer core

Inner core

The Earth's crust is, on average, between 18.9 mi. and 24.8 mi. thick beneath the continents.

Under the oceans the crust is only 3-5 mi. thick.

180 million years ago

150 million years ago

65 million years ago

Today

Q Has the Earth always looked the same as it does now?

A No. The hard outer layers are split into large pieces, or plates. These plates have been moving very slowly for millions of years, and have moved the continents around.

Q How are rocks made?

A There are three types of rock on Earth and they are made in different ways. Igneous rocks are made when hot melted material, called magma, bubbles up from the mantle and hardens. Sedimentary rocks are layers of tiny pieces of other rocks, or dead plants and animals, which gradually build up and become cemented together. Metamorphic rocks are sedimentary or igneous rocks changed by great heat and pressure.

Q What makes earthquakes happen?

A Earthquakes occur when rocks move along faults, or cracks in the Earth's outer layers. Severe earthquakes happen when plates move past each other or toward each other. The rocks at the edges of the plates rub together, making the ground shake. Sometimes the plates "stick" for a while and pressure builds up. Then they break free suddenly, causing a huge earthquake.

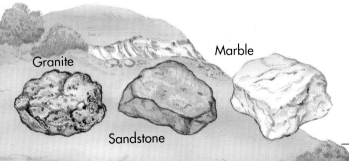

Granite

Sandstone

Marble

How many oceans are there?

The deepest scuba dive ever recorded was 452 ft. by John J. Gruener and R. Neal Watson.

Arctic
Pacific
Indian
Atlantic

There are four large oceans on Earth - the Pacific, the Atlantic, the Indian and the Arctic. There are also smaller areas of water called seas. Most seas are actually part of the oceans or are joined on to them.

Q How big are the oceans?

A Oceans and seas cover about seven-tenths of the Earth's surface. The Pacific is the largest ocean, with an area of about 64 million sq. mi., followed by the Atlantic, with about 31 million sq mi., and the Indian, at about 28 million sq. mi. The Arctic Ocean is the smallest at about 4.6 million sq. mi.

Arctic
Indian
Atlantic
Pacific

Q How were the oceans formed?

A No one is certain. Some scientists think that soon after the Earth was formed it was surrounded by thick clouds. As the Earth cooled down, rain fell and filled the hollows in the crust to form the first oceans. When the continents drifted apart, water filled the gaps they left to form the modern oceans we know today.

Q Why is seawater salty?

A The salty taste of seawater comes from minerals that have been washed into the sea by rivers. The most common mineral is salt.

Q What makes waves?

A Waves are made by wind blowing across the surface of the water. The wind pushes the water upward, making a wave crest, and then gravity pulls it back down again, into a wave trough.

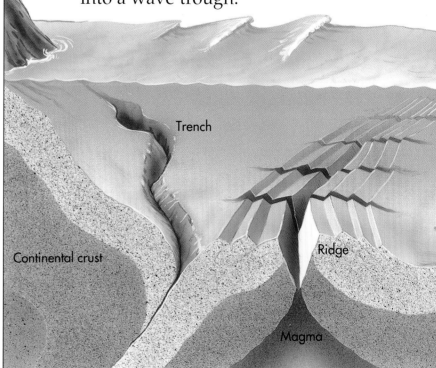

Trench

Continental crust

Ridge

Magma

Seawater contains gold - about 4 grams in every million tons of water. It also contains silver, calcium and sulphur.

The biggest wave ever recorded in the open sea was seen from a ship during a storm in 1933. It was 111 ft. high!

Q How does a coral reef grow?

A Coral reefs form in warm, shallow waters. They are made by tiny animals called polyps, which have hard, cup-shaped shells. Thousands of polyps live side by side and their shells join together to form a reef. When the polyps die, their hard shells remain. More polyps grow on top of them and gradually the reef grows.

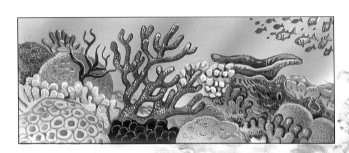

Q How are islands formed?

A Some islands are areas of land that were joined to continents long ago when the sea level was lower than it is now. The British Isles, for example, were once connected to mainland Europe. Other islands are the tops of volcanoes that rise up from the seabed. The island of Surtsey, near Iceland, was formed in this way. Between November, 1963 and June, 1967, the island rose more than 948 ft. from the ocean floor, leaving it 558 ft. above sea level.

Volcanic island

Q Is the bottom of the ocean flat?

A No. The ocean floor has many hills, valleys, deep trenches and high mountain ranges. The Mid-Atlantic Ridge, which runs for 12,428 mi. down the center of the Atlantic Ocean, is the longest mountain range in the world. The deepest point is the Marianas Trench, in the Pacific Ocean which plunges to 36,199 ft. below sea level.

Oceanic crust

How are coastlines shaped?

Coastlines are where the land meets the sea. There are many different types of coastline, ranging from sandy beaches to rocky cliffs. They are all shaped by the sea.

Q Do coastlines always stay the same shape?

A No, they are always changing. Some coastlines are being worn away by the pounding of the sea. In other places, beaches are being made.

Q Is the sand on beaches always yellow?

A No, not always. Yellow sand consists mainly of a mineral called quartz. In the Hawaiian Islands, the sand is black because it is made from black lava produced by erupting volcanoes.

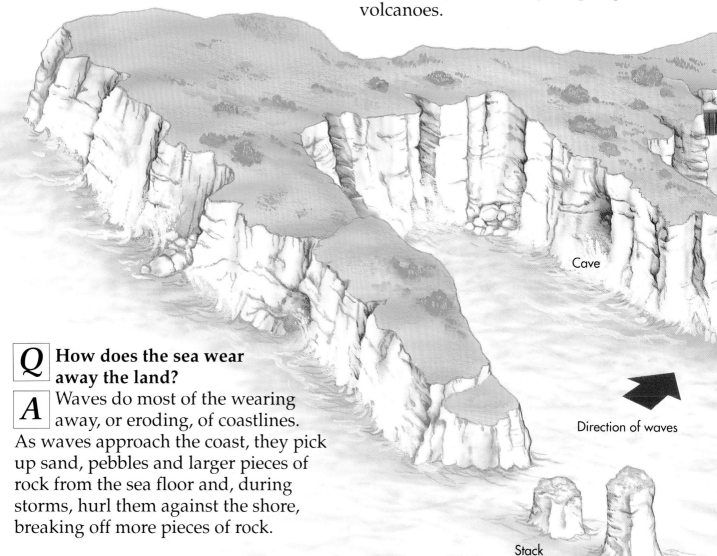

Cave

Direction of waves

Stack

Q How does the sea wear away the land?

A Waves do most of the wearing away, or eroding, of coastlines. As waves approach the coast, they pick up sand, pebbles and larger pieces of rock from the sea floor and, during storms, hurl them against the shore, breaking off more pieces of rock.

Durdle Dor in England is a good example of an arch.

A stack is a single column of rock. The Old Man of Hoy in the Orkney Islands is a famous example of a stack.

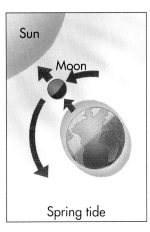

Q How are beaches made?

A Beaches form along stretches of coast that are sheltered from strong waves, such as in bays. Most beach sand is brought from inland by rivers flowing into the sea. Some is made when rocks broken off by the sea are worn down into smaller and smaller pieces. Sand is dropped along the shore by waves and it builds up into beaches.

Q Can we take back land from the sea?

A Yes. Coastal marshes and even shallow bays can be reclaimed from the sea. The first step is usually to build a protective wall to stop the sea flooding the area. Then the salt water is drained or pumped out, leaving dry land.

Protective wall

Beach

Groynes

Q Why do some beaches have groynes?

A Groynes, or breakwaters, are built stretching into the sea to stop beaches from being worn away by waves.

Q What causes high and low tides?

A Tides rise and fall twice in every 24 hours and 50 minutes. They are caused mainly by the Moon. When the Moon is overhead, its gravity pulls upward on the sea, away from the shore, causing a bulge like a large wave. This is low tide. As the Earth spins, the wave travels around the planet, causing high tides.

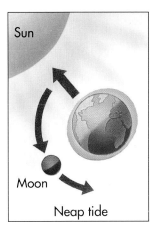

Sun

Moon

Spring tide

Sun

Moon

Neap tide

Spring and neap tides occur twice a month. They are the highest and lowest high tides.

These are the five highest mountains in the world.

Mount Everest 29,030 ft.

The first mountains may have been made soon after the Earth was formed, but they were worn away many millions of years ago. The mountains we can see today are much younger.

Q Where are the world's highest mountains?

A The world's 20 highest mountains (measured from sea level) are all in the Himalaya-Karakoram range in Asia. Mount Everest is the tallest at 29,030 ft. The highest mountain from top to bottom is the peak of Mauna Kea in the Hawaiian Islands. It rises 33,476 ft. from the ocean floor, but as over half of it is underwater only 13,797 ft. rises above sea level.

Q Do mountains always stay the same?

A No. Mountains are being worn away by rain, wind, frost and other natural forces. Some mountain ranges, such as the Alps, Himalayas and Andes, are still rising as the continental plates they rest on are pushed closer together.

Q What is a rift valley?

A Where two faults run side by side, the block of land between them may sink down to form a rift valley. The most famous example is the Great Rift Valley which runs for 3,977 mi. from Syria down through East Africa.

Q How are mountains made?

A Some mountains are volcanoes. Others are dome mountains which were pushed up by hot melted, or molten, rock rising below the surface. Some mountains formed when rocks were squeezed together and folded. Others are blocks of land forced up between huge cracks, or faults, in the Earth's surface.

Volcanic

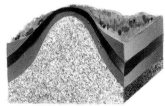

Dome

Folds

Fault

K2
28,709 ft.

Kangchenjunga
28,207 ft.

Lhotse
27,924 ft.

Makalu I
27,826 ft.

Active volcanoes around
the world.

Q **Why do high mountains have snow on top?**

A For every 3,200 ft. that you go up, the temperature of the air falls by about 13°F. The tops of high mountains are always surrounded by very cold air, even in summer.

Q **What is a volcano?**

A A volcano is a hole in the Earth's crust. When a volcano erupts, hot molten rocks from inside the Earth pour out of the hole on to the surface. Volcanoes that erupt often are called active, while those that might erupt some time in the future are said to be dormant. A volcano that has stopped erupting is said to be extinct.

Q **Where are there volcanoes?**

A There are about 1,300 active volcanoes (ones that erupt) in the world, although only about twenty to thirty erupt in any one year. Most volcanoes are in areas near the edges of the plates which make up the Earth's outer layers.

The deepest lake is Lake Baikal, in Russia, which is 6,365 ft. deep.

The water in rivers comes from rain, lakes, springs and melting ice and snow. Rivers often begin high up on mountains and run downhill to the sea. As they flow, they wear away the land to make valleys.

Q Why do rivers get larger as they flow?

A A river usually starts as a tiny trickle called a rill. It flows downhill and is joined by other rills. It gets larger and becomes a stream. Other streams flow into it to make a river. A river may have other rivers, called tributaries, flowing into it and so it gets bigger and bigger.

Q Which is the world's longest river?

A The Nile is the longest, at 4,160 mi. It flows through East Africa into the Mediterranean Sea. The river that contains the most water is the Amazon in South America. Every second it carries about 4,237,680 cu. ft. of water into the Atlantic Ocean.

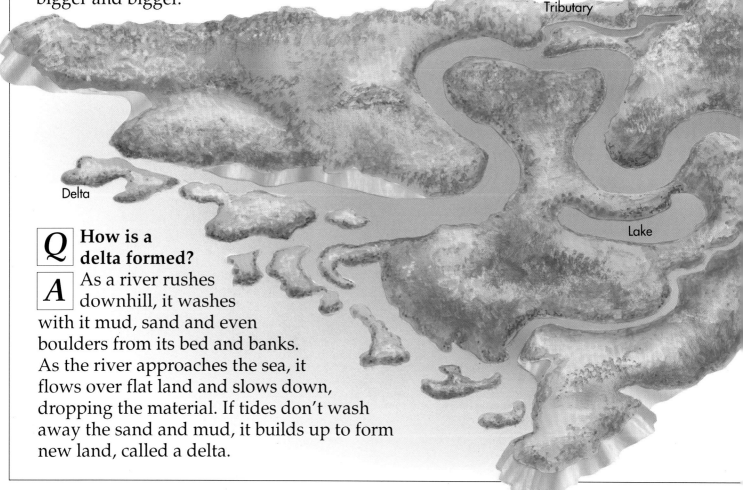

Tributary

Delta

Lake

Q How is a delta formed?

A As a river rushes downhill, it washes with it mud, sand and even boulders from its bed and banks. As the river approaches the sea, it flows over flat land and slows down, dropping the material. If tides don't wash away the sand and mud, it builds up to form new land, called a delta.

The world's largest freshwater lake is Lake Superior. It covers an area of 31,660 sq. mi. in the U.S. and Canada.

Each year the Huang He, or Yellow River, in China washes about 2,240 million tons of soil down its valley.

Q How are lakes formed?

A Most of the world's lakes are in places where glaciers carved out deep valleys in the land (see pages 18-19). Some lakes, such as Lake Tanganyika in Africa, are in deep valleys called rift valleys, which are made by huge movements in the Earth's crust. A lake may form in the craters of volcanoes, or when a river changes its course.

Q Why don't rivers run straight?

A On their way downhill to the sea, rivers travel along the easiest routes. If a river meets something in its way, such as a boulder or a hill, it simply flows around it.

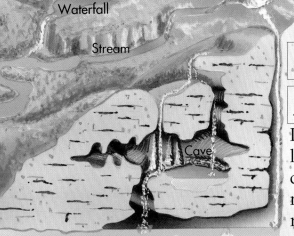

Waterfall

Stream

Cave

Underground river

Q Are there rivers under the ground?

A Yes, especially in areas where there is limestone. Rainwater can wear away limestone, making holes and caves. A river on the surface may pour down a hole in the rock and flow through the caves underground.

Q How are waterfalls made?

A Most waterfalls form where a river flows over a layer of hard rock and then over softer rock. The river wears away the softer rock faster than the hard rock, making a step. The step gradually gets deeper and the river plunges over it, creating a waterfall.

Many plants and animals have adapted to living in the deserts.

Are all deserts hot and sandy?

The saguaro cactus can grow up to 50 ft. high.

An area is called desert if, on average, it has less than ten inches of rain a year. Deserts are often in hot places, but not always. Most deserts are covered with rocks and stones rather than sand.

Q Where are the world's deserts?

A The world's largest deserts are in regions with high air pressure. Winds blow outward from these areas and moist winds from the sea very rarely blow into them. Other deserts are far from the sea. By the time winds reach them, they have lost most of their moisture. Some deserts are on the inland, sheltered sides of mountain ranges. Most of Antarctica is a frozen desert. It is a region of high pressure and little new snow falls inland.

Scorpion

Q Which is the world's driest desert?

A Many deserts go without rain for several years at a time and then have a short downpour. The driest desert is the Atacama in South America; until 1971, it had not rained there for four hundred years.

Q What is an oasis?

A An oasis is an area of land in a desert where plants are able to grow because there is water from an underground spring or a well.

Q How do sand dunes move?

A Wind blows loose sand along the ground and piles it up into hills called dunes. Sand grains are blown up one side of a dune and rolled over the top and down the steep face on the other side. Sand grains are always being blown up and over the dune, moving it across the desert.

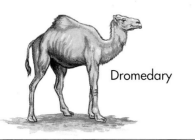
Dromedary

Cholla

Ocetillo

Kangaroo rat

Q Which is the largest desert?

A The Sahara desert, in North Africa, is the biggest. It covers an area of about 3,243,240 sq. mi. Only about a tenth of it is covered with sand; the rest is rocky.

The World's deserts

Cactus

Hot desert

Polar desert

Q Can plants and animals live in the desert?

A Yes. Plants and animals have developed special ways of living in deserts. Some plant seeds stay buried in the sand for years until it rains. Then they grow quickly and produce seeds of their own before dying off when the sand dries out. Most desert animals hide during the day and come out at night when it is cooler. Some can store water in their bodies for a long time.

Rattlesnake

Gila monster

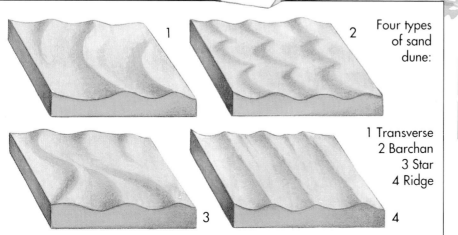

Four types of sand dune:

1 Transverse
2 Barchan
3 Star
4 Ridge

Q Do deserts change size?

A Sometimes they grow and at other times they shrink. In 1987 the Sahara spread 33 mi. south in one area, but the following year it retreated 60 mi. north.

Ice sheets can cover vast areas during an ice age.

What is a glacier?

The ice in most glaciers is between 650 and 1,312 ft. thick.

A glacier is like a river of ice. It forms when snow piles up in a hollow on a mountainside. As the snow continues to build up it gets squeezed together turning into ice. Finally, the ice spills out of the hollow and flows slowly downhill as a glacier.

Q What is an ice age?

A An ice age is a period in which temperatures are much lower than normal and ice sheets spread across large parts of the Earth. Ice ages happen every few million years. The most recent one ended about ten thousand years ago. Parts of North America and Europe were covered by ice sheets.

Q How thick is an ice sheet?

A The ice sheets that covered much of North America and Europe during the last ice age were up to 9,800 ft. thick. The ice covering the continent of Antarctica today is even thicker - 15,740 ft. in places.

A cross-section through an ice sheet.

Q Are there any glaciers in the world today?

A Yes. There are glaciers in the cold lands at the far north and south of the Earth, in northern Canada, Greenland and Antarctica, for example. Glaciers also slide down valleys in high mountain ranges, such as the Himalayas, the Rockies and the Alps.

Q Why do glaciers stop moving?

A As a glacier slides down a mountain it moves into warmer regions and begins to melt. Eventually it gets to a point where it is melting at the bottom at the same rate that it is being fed by fresh snow at the top. In the cold north and south, glaciers may flow straight into the sea.

A close-up view of a melting glacier.

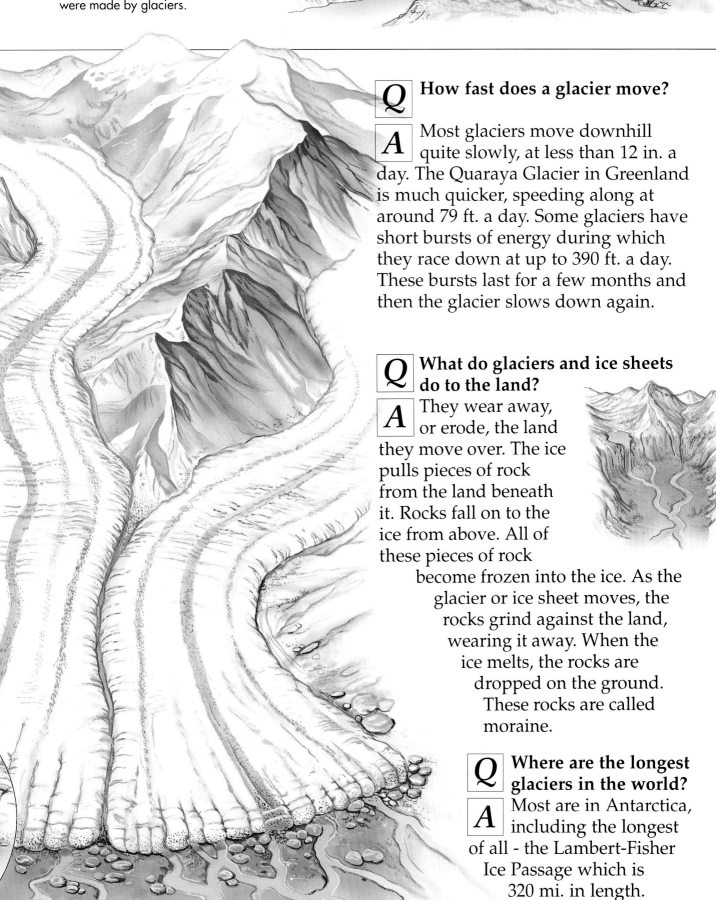

The deep sea inlets, or fjords, along the coasts of Scandinavia were made by glaciers.

Q How fast does a glacier move?

A Most glaciers move downhill quite slowly, at less than 12 in. a day. The Quaraya Glacier in Greenland is much quicker, speeding along at around 79 ft. a day. Some glaciers have short bursts of energy during which they race down at up to 390 ft. a day. These bursts last for a few months and then the glacier slows down again.

Q What do glaciers and ice sheets do to the land?

A They wear away, or erode, the land they move over. The ice pulls pieces of rock from the land beneath it. Rocks fall on to the ice from above. All of these pieces of rock become frozen into the ice. As the glacier or ice sheet moves, the rocks grind against the land, wearing it away. When the ice melts, the rocks are dropped on the ground. These rocks are called moraine.

Q Where are the longest glaciers in the world?

A Most are in Antarctica, including the longest of all - the Lambert-Fisher Ice Passage which is 320 mi. in length.

20

In 1958 the U.S. Navy submarine *Nautilus* actually traveled underneath the North Pole.

Antarctica contains ninety percent of the world's ice.

The polar regions are at the far north and south of the Earth. The Arctic lies inside the Arctic Circle, an imaginary line around the North Pole. The Antarctic is a continent that surrounds the South Pole.

Arctic circle

North Pole

Arctic wildlife

Q How much of the world is covered with ice?

A Ice covers more than one-tenth of the Earth's land surface. The Greenland ice sheet covers about 695,000 sq. mi. while the ice sheet in Antarctica is over 5,000,000 sq. mi. About three-quarters of all the world's fresh water is frozen in ice sheets and glaciers.

Q What are the North and South Poles?

A The Poles are the places on the Earth which are farthest north and south. If you stood at the North Pole, which ever way you walked you would be going south.

Q What is underneath the polar ice?

A Beneath the ice in Antarctica lies a vast continent, with high mountains and deep valleys. The ice around the North Pole in the Arctic is not resting on land but simply floating on the sea.

Here are three of the many different shapes of iceberg that can be found at the Poles.

Slab

Jagged

Pyramidal

Q What is an iceberg?

A Where ice sheets and glaciers flow down into the sea, huge chunks of ice - icebergs - can break off and float away. The tallest one ever seen stood 548 ft. above the water and extended more than a mile below the surface. The largest iceberg ever seen was near Antarctica in 1956. It covered an area of more than 11,970 sq. mi.

Q Which is colder, the Arctic or the Antarctic?

A The Antarctic gets far colder than the Arctic. The coldest place of all is the Vostok Base in Antarctica, close to the South Pole. In July, 1983, an air temperature of -243.7°F was recorded there.

Q Is it always snowing in the Antarctic?

A No, the Antarctic actually gets very little snow, especially in the center. The air over the continent is very cold and also very dry, and only about 2 in. of snow falls in a year. Close to the coasts, where it is not so cold and there is more moisture in the air, much more snow falls.

Q Does anyone live in Antarctica?

A No one lives there permanently. There are a number of research stations where scientists live for months at a time and study the area. Away from the coast, Antarctica has almost no life at all other than one type of tiny mite, several mosses and two species of flowering plant.

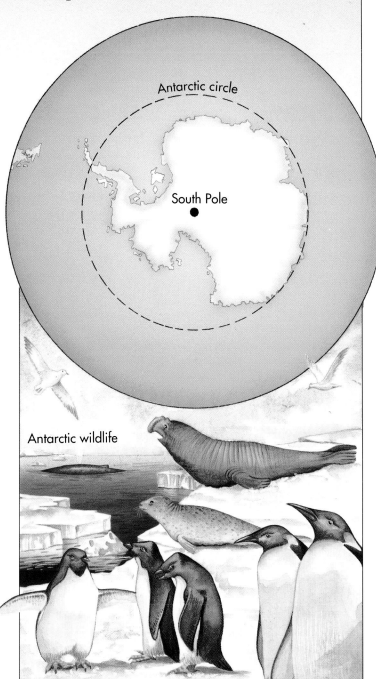

Antarctic circle

South Pole

Antarctic wildlife

Plants give off oxygen and make the atmosphere capable of supporting animal life.

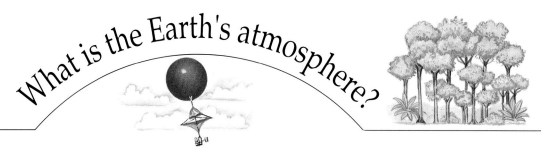

What is the Earth's atmosphere?

The atmosphere is the layer of gas that surrounds the Earth. The atmosphere is made up of oxygen (which we all need to breathe), nitrogen, water vapor and tiny amounts of other gases.

Q How many layers are there in the atmosphere?

A The atmosphere is divided into five layers - the troposphere, stratosphere, mesosphere, thermosphere and exosphere.

Q Why doesn't the atmosphere float away into space?

A The force of gravity pulls the gasses of the atmosphere towards the Earth and stops them from drifting away. More than three-quarters of the gasses are squashed into the troposphere, and there are fewer and fewer the higher you go. In the exosphere there are hardly any gasses at all.

Q Does the atmosphere get colder the higher up you go?

A Yes and no. It gets colder as you climb through the troposphere, then warmer in the stratosphere, colder again in the mesosphere and then warmer again in the thermosphere and exosphere.

Exosphere

Thermosphere

Mesosphere

Stratosphere

Troposphere

Q Does air weigh anything?

A Yes, scientists think that the weight of all the air around the Earth is about 5,200 million million tons.

Oxygen only makes up about 20 per cent of the atmosphere. Nitrogen makes up 78 per cent of the atmosphere.

These are scientists' models of oxygen and nitrogen.

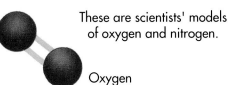

Nitrogen

Oxygen

Q **Where is the ozone layer?**

A It is in the stratosphere, between 9 and 17 miles above the Earth. It consists of a thin layer of ozone gas, which is a type of oxygen. Ozone absorbs much of the harmful ultraviolet radiation from the Sun, preventing it from reaching the Earth. Without the ozone layer, all living things on Earth would be killed by the ultraviolet rays.

The ozone layer is an invisible barrier protecting the Earth.

Q **What is atmospheric pressure?**

A Because air has weight, it presses on the Earth and everything on it, including us. The more air there is above us, the more it presses on every cubic inch - or the higher the atmospheric pressure. So, the higher up we go and the less air there is above us, the lower the pressure is.

Q **Why does air move about in the atmosphere?**

A Moving air in the atmosphere is better known as wind. It is caused by differences in air temperature. Warm air weighs less than cold air, and so it rises. Cooler air then rushes in underneath it to take its place, creating winds.

The winds in a hurricane are thought to reach about 280 miles per hour.

Where does weather happen?

All of the world's weather takes place in the lowest layer of the atmosphere - the troposphere. That is where the Sun's heat causes air to rise, cool and sink down again. The movement of the air creates areas of high and low pressure and winds.

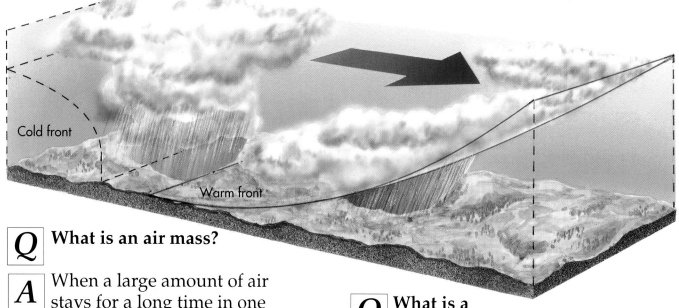

Cold front

Warm front

Q What is an air mass?

A When a large amount of air stays for a long time in one place, it forms an air mass. They are warm or cold, depending upon the temperature of the land or sea below. When air masses start to move, they bring changes in the weather.

Q What is a weather front?

A A front is the boundary between a cold air mass and a warm air mass; it is where most weather changes happen.

Q Do all clouds bring rain or snow?

A No. All clouds are made of water vapor, but don't always bring rain or snow. There are ten different types of cloud and they can help to tell us what sort of weather to expect.

Cirrus - high, wispy clouds (possible rain)

Cirrostratus - thin, whitish clouds (possible rain or snow)

Cirrocumulus - thin, white ripples (changing weather)

Altostratus - grey or bluish layer of cloud (fine weather)

Altocumulus - like altostratus, but lower in the sky and fluffier (changing weather)

Cumulus - fluffy, white clouds (good weather)

Nowadays every hurricane is given a name to identify it. The names are chosen alphabetically and change from female names to male names for each new storm.

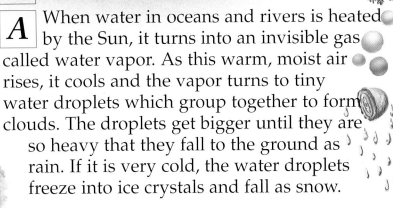

Q Why do hurricanes happen?

A Hurricanes are very powerful storms that happen in warm areas of the world. They start near the Equator when warm, wet air over the sea rises and forms giant columns of cloud full of water vapor. Cold air rushes in below the rising warm air and begins to spiral around at up to 187 miles per hour. When a hurricane reaches land, it can create enormous destruction.

Q Where is the snowiest place on Earth?

A The most snow in a year - 102 ft - fell on Paradise in Washington State, USA in 1971-72.

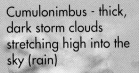

Q What makes rain and snow?

A When water in oceans and rivers is heated by the Sun, it turns into an invisible gas called water vapor. As this warm, moist air rises, it cools and the vapor turns to tiny water droplets which group together to form clouds. The droplets get bigger until they are so heavy that they fall to the ground as rain. If it is very cold, the water droplets freeze into ice crystals and fall as snow.

Q What causes thunder and lightning?

A When warm, moist air rises very quickly, deep cumulo-nimbus storm clouds form. Ice crystals and water droplets whirl around inside the clouds and bump into each other, making tiny electric charges. The charges build up until huge electric sparks flash from cloud to cloud or down to the ground and back. The lightning flash heats up the air as it passes. The air expands very rapidly and makes the booming noise we call thunder.

Cumulonimbus - thick, dark storm clouds stretching high into the sky (rain)

Stratocumulus - uneven patchy clouds (dry weather)

Stratus - low layer of grey cloud (rain or snow)

Nimbostratus - dark grey layer (rain or snow)

Siberia, in Russia, has the greatest range of temperatures during a year, from -191°F to 102°F.

What is climate?

The type of weather that a place usually has from year to year is called its climate. Some places are warm all year round while others are always very cold. In some areas the temperature changes from season to season.

Q Why do some places have hotter climates than others?

A The hottest climates are usually in places closest to the Equator. The Sun's rays are more concentrated there than they are further north or south, where the Earth's rounded shape spreads the Sun's rays over a larger area.

Q How many types of climate are there?

A The world is divided into four main types of climate: polar, temperate, subtropical and tropical. Some areas, such as deserts and mountains, have their own special types of climate.

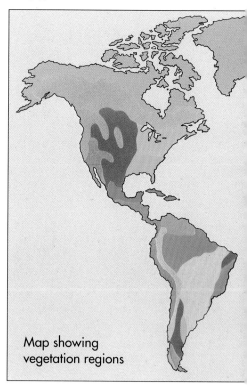

Map showing vegetation regions

Q What things can affect climate?

A The climate of a particular place is affected by a number of things. Places near the sea have milder climates than those far inland, which often have very hot summers and freezing winters. Ocean currents can make climates warmer or colder. Mountains have colder climates than the lowlands around them.

Q Do cities have their own climates?

A Yes, cities are often warmer than the surrounding areas. Concrete buildings absorb heat from the Sun during the day and release it at night, making the air in the city warmer.

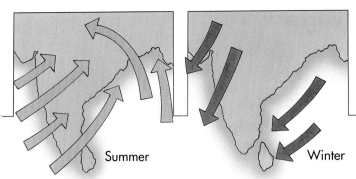

Summer   Winter

The seasons come at different times of the year in different places.

When it is summer in Europe it is winter in Australia.

Q What is a monsoon?

A A monsoon is a type of wind. In some places near the Equator, air over the land heats up in summer and rises. Cool, wet air from the ocean blows in to take its place, bringing heavy rain. In winter the winds reverse and dry air blows from the land towards the sea.

Q Does climate affect which plants grow where?

A Yes. Different types of plant prefer different climates. For example, rainforests grow in the hot, wet areas around the Equator, and coniferous forests grow in the cold areas in the north. Farther north, where it is even colder, only mosses and tiny flowers can survive.

Polar and tundra

Taiga and mountains

Temperate forest

Dry grassland

Mediterranean

Desert and semi-desert

Tropical grassland

Rainforest

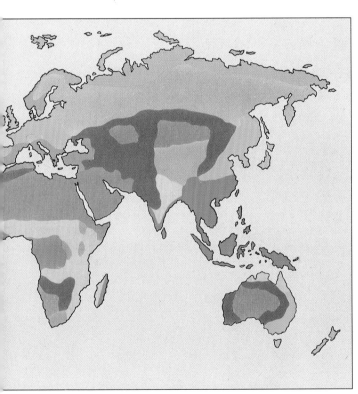

Q Why do we have seasons?

A As the Earth moves around the Sun, different parts of it are tilted towards the Sun for a few months at a time. The part leaning towards the Sun has summer, while the part leaning away has winter.

More than 20 percent of the world's oil comes from the rocks beneath the sea.

The first large scale nuclear plant was built in 1956.

Almost everything we use, such as the fuel we burn, the wood and stone we use for building and the food we eat, comes from the Earth in some way. These things are called natural resources.

Q What are fossil fuels?

A They are fuels, such as coal, oil and gas, that were formed from the remains of animals and plants that lived and died millions of years ago.

Q What are renewable energy sources?

A When fossil fuels are used they can't be renewed, or made again. Some energy sources will never run out. They are called renewable sources and include running water, the wind, energy from the Sun and even heat from the rocks deep inside the Earth.

Q How is nuclear power produced?

A Everything in the Universe is made of tiny particles, called atoms. When atoms are broken apart they give off heat, which can be used to produce electricity. This is done inside nuclear power stations. The atoms most often used are uranium.

Hydroelectric dams produce electricity from running water.

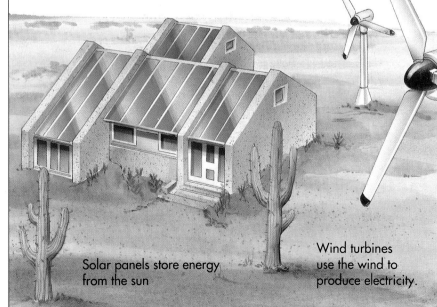

Solar panels store energy from the sun

Wind turbines use the wind to produce electricity.

Geothermal plants use energy from rocks below the ground.

There are about 70,000 different varieties of rice in the world.

Q Which food crops are the most important?

A Cereals, or grain crops, are used to feed more people in the world than any other type. Rice and wheat are the most important grain crops.

Q What are staple foods?

A Your staple food is the one that you eat most often. In Asia, for example, people eat more rice than anything else. Staple foods in Europe and North America include bread, potatoes and pasta.

Q What is a mineral?

A Minerals are natural, non-living substances that are found in the Earth. There are about 3,000 types of mineral, each made from its own special combination of atoms. Some minerals, such as diamonds, rubies and emeralds are very rare and valuable. Others, like quartz, are extremely common.

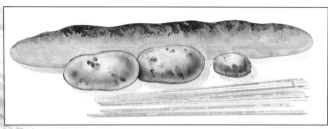

Copper

Quartz

Ruby

Diamond

Q How many fish can be caught in one go?

A In 1986 a Norwegian fishing boat hauled in a single catch weighing more than 2,300 tons. It was estimated to contain about 120 million fish.

Q Do the oceans have resources?

A Yes. The rocks of the sea-bed contain minerals, including oil and gas, and metals. We catch fish to eat in the oceans, and the seawater itself can be made into water for drinking. Waves and tides are also renewable energy sources.

The U.S. has less than five percent of the world's people, but uses twenty-nine percent of the Earth's gasoline.

Are we damaging the planet?

We are damaging the Earth in a number of ways - by polluting the air, soil and water, by destroying the places where plants and animals live, and by using up its natural resources.

Q What is pollution?

A Pollution means spoiling our environment by putting into the air, water or land materials that will harm it. Humans have always polluted the planet with smoke, trash and other things. Pollution is now very serious because there are many more humans than ever before.

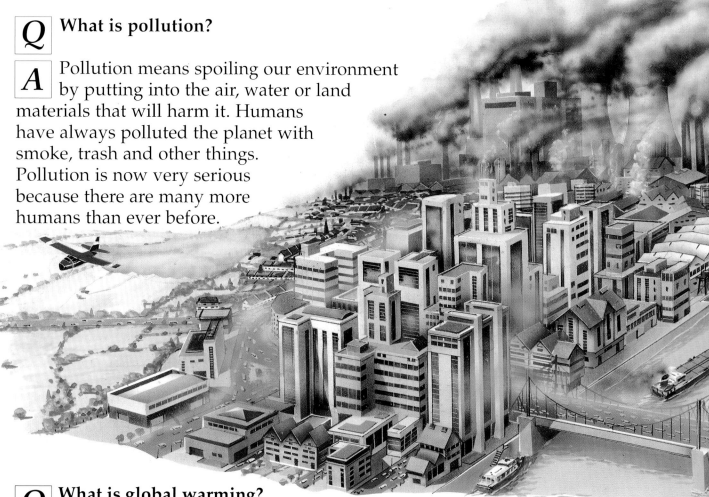

Q What is global warming?

A Some gases in the atmosphere, such as carbon dioxide, help to trap in heat from the Sun. Factories, cars and power stations produce a lot more of these gases, and so more heat is being trapped in. If the Earth keeps getting warmer, the Antarctic ice sheet may begin to melt and the sea level would rise, flooding some low coasts.

Q Where does our trash go?

A Some is burned, which causes air pollution and adds to global warming. Most is either buried in holes in the ground or dumped at sea, causing more pollution. Little is reused or recycled, even though glass, metals, paper and plastics can all be treated and reused.

There are between five and eight million different species of plants and animals in the world. So far, only 1.6 million have been identified and many of the rest may become extinct even before we discover them.

Q How does rain become acid?

A All rainwater is very slightly acidic. When fossil fuels are burned - in car engines, factories and power stations - chemicals are produced which make the moisture in the air much more acidic. Eventually, this moisture falls to the ground as acid rain.

Q Are there holes in the ozone layer?

A The ozone layer protects us from harmful ultraviolet radiation from the Sun. The ozone is being destroyed by chemicals called CFCs, which are used in aerosol cans, coolants, and some plastics. So far, there aren't any actual holes in the ozone layer, but it has become thin in places. Many countries have now stopped using CFCs.

A satellite view of the ozone layer. The pink color shows the thin areas.

Q Why do people cut down rainforests?

A The world's tropical rainforests are cut down to make grazing land for cattle and to use certain hardwood trees for making furniture. The forests are being destroyed at a rate of about 9 sq. mi. every hour. If this continues, by the year 2050 there will be none left.

Q Does recycling paper save rainforests?

A No, because the trees used for making paper are specially grown. Recycling paper makes sense because it uses far less energy and water than making new paper from trees. It also means that we throw less away, which reduces pollution.

Index